j523.7 RAT
Rathburn, Betsy
The sun.
$21.95

11/2018

SPACE SCIENCE

THE SUN

BY BETSY RATHBURN

BELLWETHER MEDIA • MINNEAPOLIS, MN

Are you ready to take it to the extreme? Torque books thrust you into the action-packed world of sports, vehicles, mystery, and adventure. These books may include dirt, smoke, fire, and chilling tales. **WARNING**: read at your own risk.

This edition first published in 2019 by Bellwether Media, Inc.

No part of this publication may be reproduced in whole or in part without written permission of the publisher.
For information regarding permission, write to Bellwether Media, Inc., Attention: Permissions Department,
6012 Blue Circle Drive, Minnetonka, MN 55343.

Library of Congress Cataloging-in-Publication Data

Names: Rathburn, Betsy, author.
Title: The Sun / by Betsy Rathburn.
Description: Minneapolis, MN : Bellwether Media, Inc., [2019] | Series: Torque: Space Science | Audience: Ages 7-12. | Includes bibliographical references and index.
Identifiers: LCCN 2018001115 (print) | LCCN 2018008521 (ebook) | ISBN 9781681036038 (ebook) | ISBN 9781626178625 (hardcover : alk. paper)
Subjects: LCSH: Sun–Juvenile literature.
Classification: LCC QB521.5 (ebook) | LCC QB521.5 .R3485 2019 (print) | DDC 523.7–dc23
LC record available at https://lccn.loc.gov/2018001115

Text copyright © 2019 by Bellwether Media, Inc. TORQUE and associated logos are trademarks and/or registered trademarks of Bellwether Media, Inc. SCHOLASTIC, CHILDREN'S PRESS, and associated logos are trademarks and/or registered trademarks of Scholastic Inc., 557 Broadway, New York, NY 10012.

Editor: Rebecca Sabelko   Designer: Andrea Schneider

Printed in the United States of America, North Mankato, MN.

TABLE OF CONTENTS

Show Stopper 4
What Is the Sun? 6
What Is the Sun Made Of? 10
Where Is the Sun? 16
Why Do We Study the Sun? 18
Glossary 22
To Learn More 23
Index 24

It is August 21, 2017. Millions of people across the United States gather to witness a total **solar eclipse**. They turn their faces to the sky. The show is about to start.

As minutes tick by, the Moon slowly covers the Sun. The crowd gasps as the Sun goes dark. All that remains is a beautiful ring of light!

AUGUST 21, 2017
SOLAR ECLIPSE

WHAT IS THE SUN?

The Sun is a medium-sized star known as a **yellow dwarf**. It is more than 2.7 million miles (4.3 million kilometers) around. Earth could fit inside it more than one million times!

The Earth's **atmosphere** changes the Sun's appearance. From Earth, the Sun looks yellow. But it is really white!

FUN FACT

A NEIGHBORING LIGHT

Besides the Sun, Earth's next-closest star is Proxima Centauri. This star is trillions of miles from Earth!

The Sun is the brightest star in Earth's sky. It provides light and warmth from about 93 million miles (149 million kilometers) away.

DISTANCE BETWEEN EARTH AND THE SUN

EARTH TO SUN = 92,900,000 MILES
(149,500,000 KILOMETERS)

EARTH TO MOON = 238,855 MILES
(384,400 KILOMETERS)

Like other yellow dwarf stars, the Sun is very hot! It has a surface temperature of about 10,000 degrees Fahrenheit (5,500 degrees Celsius).

WHAT IS THE SUN MADE OF?

The Sun is about 4.6 billion years old. It formed from a spinning cloud of dust and gas called a **solar nebula**.

Gravity flattened the cloud into a disk. Then, it pulled the materials to the center of the disk. This formed the Sun.

11

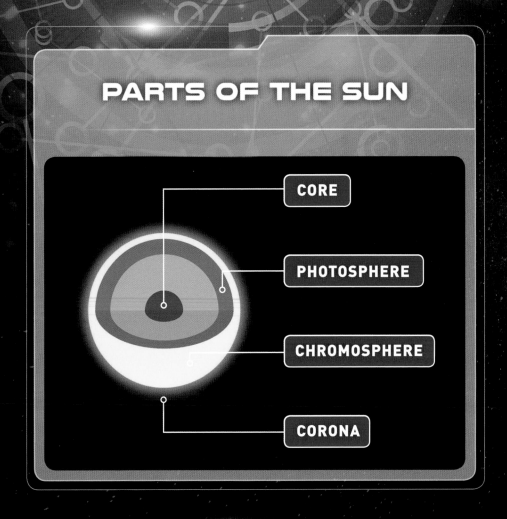

Today, the Sun is made of layers of hot gases. The Sun's **core** releases huge amounts of heat and energy. This energy travels outward to the Sun's **photosphere**.

SUNSPOT

Sometimes, dark patches are visible on parts of the photosphere. These **sunspots** are caused by activity deep within the Sun. The biggest ones are thousands of miles across!

Beyond the photosphere are the **chromosphere** and **corona**. The chromosphere can reach almost 14,000 degrees Fahrenheit (7,700 degrees Celsius). It gets hotter as it gets further from the Sun's core.

CORONA

The corona is the Sun's outermost layer. It is only visible during a total solar eclipse. When this occurs, the corona looks like a bright ring of light.

WHERE IS THE SUN?

The Sun rotates at the center of the solar system. Because of its huge size, the Sun's **equator** spins faster than the **poles**. It takes up to 36 days for the poles to rotate once!

As the Sun rotates, many objects circle around it. Earth takes 365 days to make one trip around the Sun. Other objects may take thousands of years!

FUN FACT

SPEEDY TRAVELER
Mercury circles the Sun faster than any other planet. It makes one trip in just 88 days!

WHY DO WE STUDY THE SUN?

The Sun is Earth's closest star. Scientists study it to uncover information about faraway stars.

They use special instruments that detect the Sun's energy. Their findings help them study other solar systems in the Milky Way **galaxy**. They use what they learn to study the whole universe!

FUN FACT

TOUCH THE SUN

In 2018, the Parker Solar Probe is planned to launch. It will fly closer to the Sun than any other object!

Studying the Sun helps people on Earth. The Sun's **solar flares** can disrupt our daily lives. They can make cell phones and TVs stop working.

Scientists use **satellites** to predict solar flares. They use this information to protect expensive tools. Studying the Sun helps people stay in touch!

SOLAR FLARE

GLOSSARY

atmosphere—the gases that surround Earth and other planets

chromosphere—along with the corona, the outer part of the Sun's atmosphere

core—the innermost part of the Sun

corona—the outermost part of the Sun's atmosphere

equator—an imaginary circle around a star or planet that is an equal distance from each pole

galaxy—a group of stars and solar systems

gravity—the force that pulls objects toward one another

photosphere—the Sun's surface

poles—either end of a planet or star; every planet or star has two poles.

satellites—machines that are sent into space to circle around planets, moons, stars, and other objects

solar eclipse—an event in which the Moon moves between Earth and the Sun

solar flares—sudden bursts of energy from the Sun's surface

solar nebula—a huge cloud of dust and gas from which stars form

sunspots—dark spots that appear on the Sun's surface

yellow dwarf—a type of white or yellow star with a temperature between 9,000 and 10,000 degrees Fahrenheit (4,900 and 5,500 degrees Celsius)

TO LEARN MORE

AT THE LIBRARY

Aguilar, David A. *Seven Wonders of the Solar System*. New York, N.Y.: Viking, 2017.

Green, Jen. *The Sun and Our Solar System*. North Mankato, Minn.: Capstone, 2018.

Rogers, Kate. *Exploring the Sun*. New York, N.Y.: KidHaven Publishing, 2018.

ON THE WEB

Learning more about the Sun is as easy as 1, 2, 3.

1. Go to www.factsurfer.com

2. Enter "Sun" into the search box.

3. Click the "Surf" button and you will see a list of related web sites.

With factsurfer.com, finding more information is just a click away.

INDEX

atmosphere, 6
chromosphere, 14
colors, 6
core, 12, 14
corona, 14, 15
Earth, 6, 8, 9, 16, 18, 20
equator, 16
gravity, 10
light, 4, 8, 15
Milky Way galaxy, 18, 19
Moon, 4
photosphere, 12, 13, 14
poles, 16
satellites, 20
scientists, 18, 20
size, 6, 16
solar eclipse, 4, 15
solar flares, 20, 21
solar nebula, 10
solar system, 16, 18
star, 6, 8, 9, 18
sunspots, 13
temperature, 9, 14
United States, 4
universe, 18
yellow dwarf, 6, 9

The images in this book are reproduced through the courtesy of: Milissa4like, front cover, pp. 3, 4, 6, 8, 10, 12, 14, 16, 18, 20, 23 (graphic); Markus Gann, front cover (Sun), p. 2; Alan Uster, front cover, pp. 2-3 (Earth/Moon); Newcastle, p. 4 (inset); James Kirkikis, pp. 4-5; Vadim Sadovski, pp. 6-7, 8-9, 14-15, 16-17; Golden Sikorka, p. 9 (inset); HomeArt, pp. 10-11; Elegant Solution, p. 12 (inset); Juan Gaertner, pp. 12-13; Sunti, pp. 18-19; tose, p. 19 (inset); solar seven, pp. 20-21.